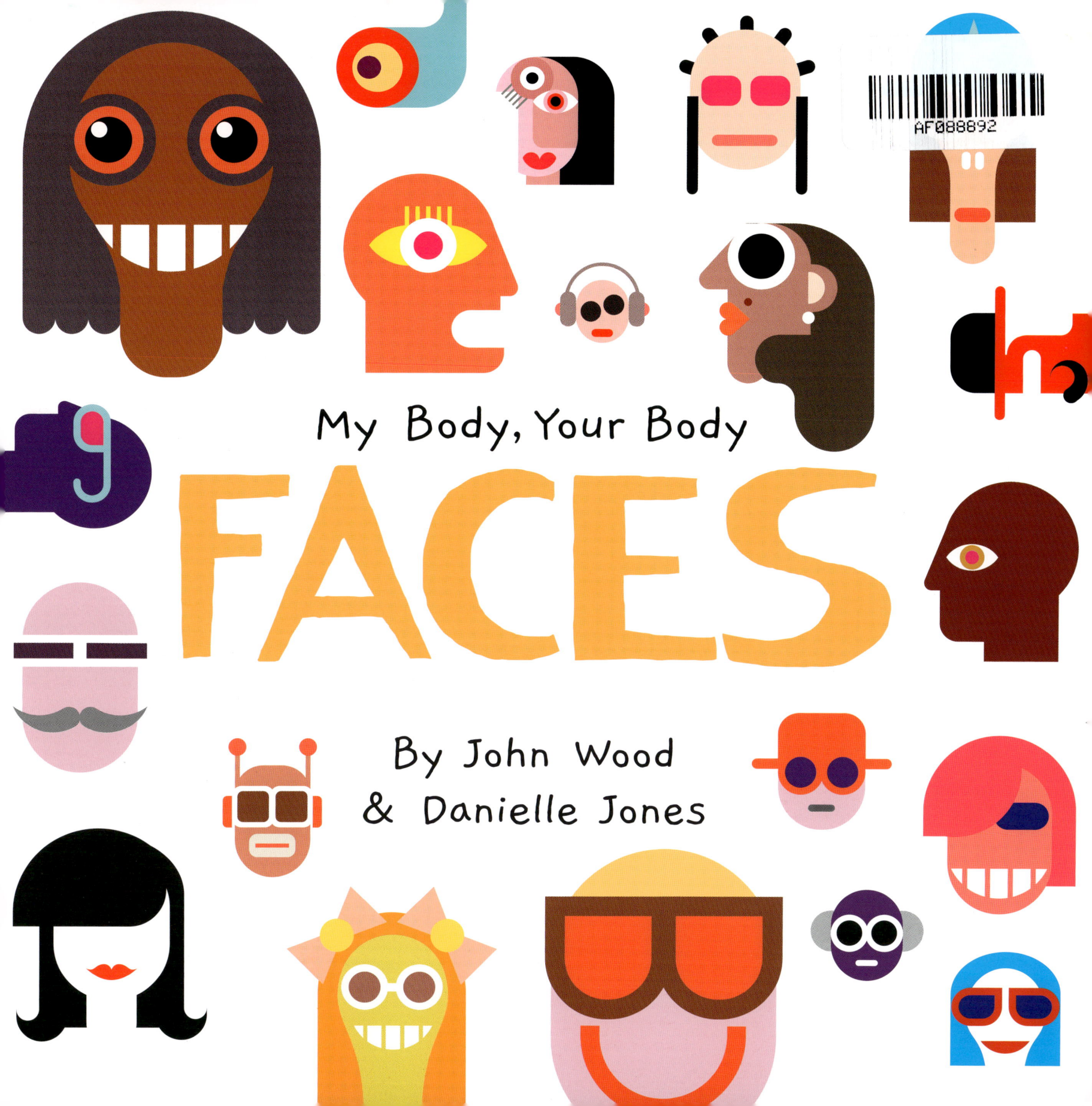

BookLife
PUBLISHING

©This edition published in 2023. First published in 2019.
BookLife Publishing Ltd.
King's Lynn, Norfolk, PE30 4LS, UK

All rights reserved. Printed in China.
A catalogue record for this book is available
from the British Library.

HB ISBN: 978-1-78637-740-1
PB ISBN: 978-1-80505-376-7

Written by: John Wood

Edited by: Madeline Tyler

Designed by: Danielle Jones

All facts, statistics, web addresses and URLs in this book were verified as valid and accurate at time of writing. No responsibility for any changes to external websites or references can be accepted by either the author or publisher.

All images are courtesy of danjazzia via Shutterstock.com, unless otherwise specified. With thanks to Getty Images, Thinkstock Photo and iStockphoto. Additional illustrations by Danielle Jones.

This is my face.

And that is your face.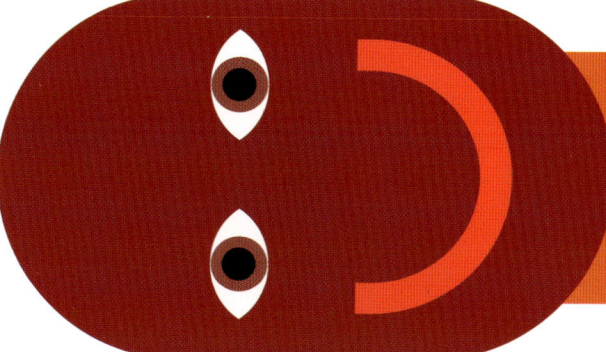

WE ALL HAVE FACES.

Her face is kind. It is warm.
It is **WIDE**.

His face looks small when he turns to one side.

His chin is thin.
It is pointy and long.

Her nose is BIG.
It is lovely and strong.

7

This face has dimples  and this face has none.

This face is glowing
and round like the Sun.

9

He starts to **frown** when he thinks for a while.

I can see all of their teeth when they **smile**.

These eyes are brown.
They get **WIDE** when she speaks.

This man has **crinkles** and lines round his cheeks.

13

Gran has BIG ears,
but she always says

"WHAT?"

Where is this face?

It is covered with hair!

This person has a nice ring in their nose.

She has **thick** glasses as red as a rose.

She puts on make-up. The colours are BRIGHT.

It takes a **long** time to get it just right.

Look at his eyelashes.
They are so long.

Her mouth gets WIDE when she sings us a song.

We would go on.
Oh, if only we could!
Faces are **different** and **lovely**
and **good**.